CONTENTS

Boldface words appear in the glossary.

Amazing Ants

What's that critter on the floor?

He's not alone—there are some more!

These busy ants can form a **mob**

when they have an important job!

Now lean in close to **inspect**

this tiny six-legged **insect**.

Lots and Lots of Ants!

How many ants live on Earth? A million?

Scientists say there are 20 **quadrillion**!

More than 12,000 kinds of ants exist.

Carpenter and fire ants are on the list.

Some are black, others brown or red.

They all have **antennae** on their head!

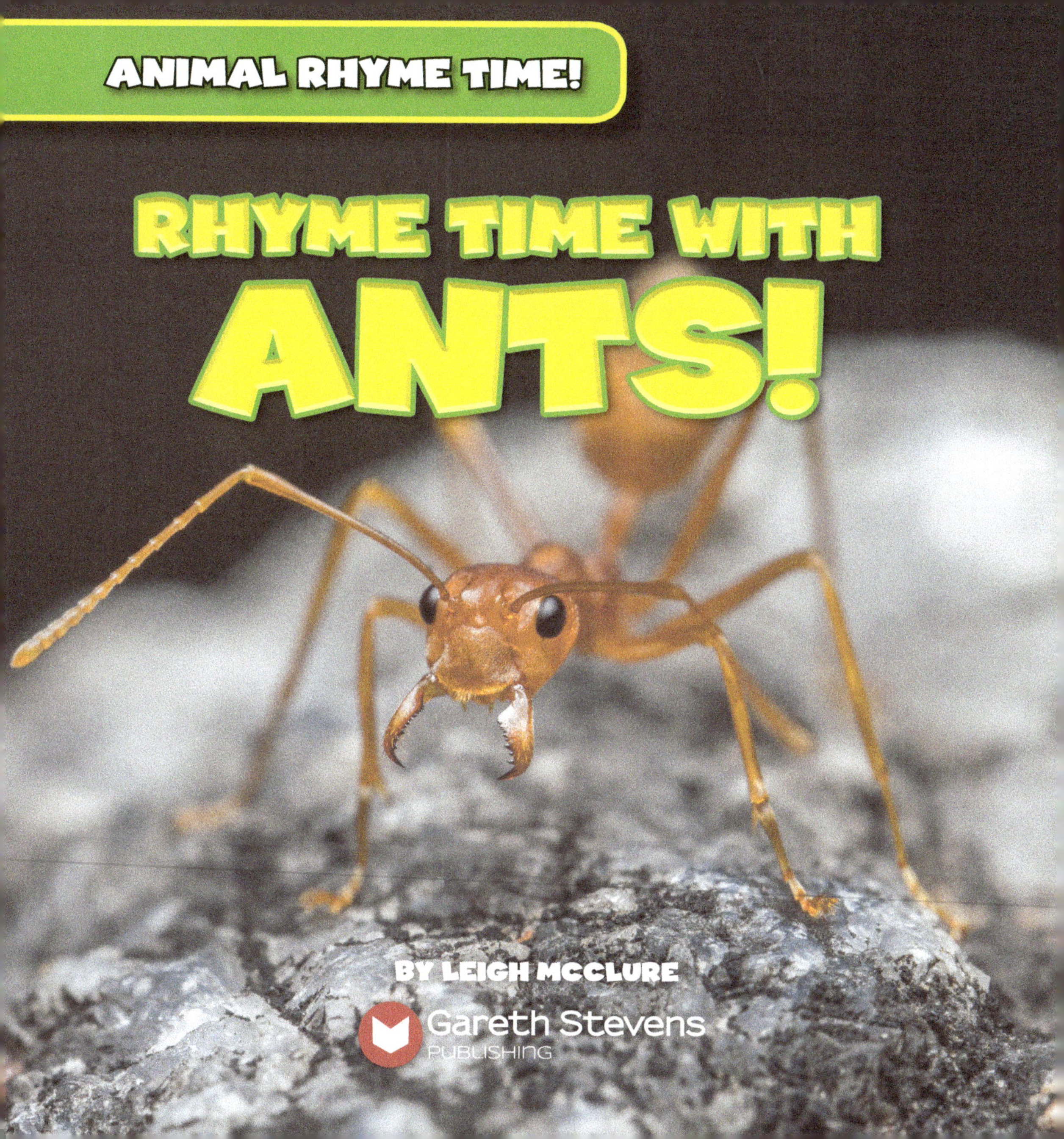

ANIMAL RHYME TIME!

RHYME TIME WITH ANTS!

BY LEIGH MCCLURE

Gareth Stevens PUBLISHING

Please visit our website, www.garethstevens.com. For a free color catalog of all our high-quality books, call toll free 1-800-542-2595 or fax 1-877-542-2596.

Library of Congress Cataloging-in-Publication Data

Names: McClure, Leigh, author.
Title: Rhyme time with ants! / Leigh McClure.
Description: Buffalo, NY : Gareth Stevens Publishing, [2024] | Series: Animal rhyme time! | Includes index.
Identifiers: LCCN 2022047253 (print) | LCCN 2022047254 (ebook) | ISBN 9781538285275 (library binding) | ISBN 9781538285268 (paperback) | ISBN 9781538285282 (ebook)
Subjects: LCSH: Ants–Juvenile literature. | Ants–Behavior–Juvenile literature.
Classification: LCC QL568.F7 M3847 2024 (print) | LCC QL568.F7 (ebook) | DDC 595.79/615–dc23/eng/20221021
LC record available at https://lccn.loc.gov/2022047253
LC ebook record available at https://lccn.loc.gov/2022047254

Published in 2024 by
Gareth Stevens Publishing
2544 Clinton Street
Buffalo, NY 14224

Designer: Andrea Davison-Bartolotta
Editor: Caitie McAneney

Photo credits: Cover Rachata Kietsirikul/Shutterstock.com; p. 5 Lapha.R/Shutterstock.com; p. 7 Imagefoto55/Shutterstock.com; p. 9 (inset) File:Dinoponera quadriceps MHNT.jpg/Wikimedia Commons; p. 9 (main) Sergei Loginov/Shutterstock.com; p. 11 Andrey Pavlov/Shutterstock.com; p. 13 GR92100/Shutterstock.com; p. 15 Kevin Wells Photography/Shutterstock.com; p. 17 Pavel Krasensky/Shutterstock.com; p. 19 (inset) Robert Gregory Griffeth/Shutterstock.com; p. 19 (main) Lidok_L/Shutterstock.com; p. 21 frank60/Shutterstock.com.

Printed in the United States of America

CPSIA compliance information: Batch #CS24GS: For further information contact Gareth Stevens, New York, New York at 1-800-542-2595.

Around the World

Does Antarctica have ants on the ground?

That's the only **continent** ants aren't found!

Ants enjoy places that are warm and wet,

plus cold or hot **climates**, Brazil to Tibet.

The largest kind is almost 2 inches long.

This ant is found in the Amazon!

GIANT
AMAZONIAN ANT

Teamwork!

The greatest thing about ants, it seems,

is that they're good at working in teams!

Ant groups are called colonies.

Queen ants rule nests with ease.

Worker ants gather food for all.

Teams are big, though ants are small!

Ants on the Move!

Ants hurry, running to and fro.

Tiny legs keep them on the go!

Some gather food. Their job is big!

Other ants carry leaves and twigs.

Leafcutter ants are farmers among us,

planting leaf bits to make a **fungus**!

Strong Insects!

The strongest creature relative to size:

tiny ants win the prize!

Another reason why ants are great—

some carry 50 times their weight!

For heavy lifting, ants ask their friends.

With big jobs, the teamwork never ends.

Long Live Ants!

Ants start as eggs, tiny as can be.

Then they grow into wiggly **larvae**.

Adult females keep the nest clean.

Adult males fly to the queen.

Queens lay eggs, growing larger than the rest.

Some queens live up to 30 years in their nest!

QUEEN ANT

Ants at Home

Do all ants live underground?

Some dig and some build mounds.

Some live in leaf litter and rotting wood.

Some think life in a house sounds good!

Ants build tunnels with rooms inside.

The largest nest was 3,700 miles wide!

ANT MOUND

Feelin' It

Ants have no ears, sometimes no eyes!

How do these tiny insects get by?

They feel **vibrations** through their feet.

Antennae "smell" and touch
when they meet.

Ants use teamwork as a tool.

With these special powers, ants rule!

GLOSSARY

antenna: A feeler on the head of some animals. The plural of "antenna" is "antennae."

climate: The average weather conditions of a place over a period of time.

continent: One of Earth's seven great landmasses.

fungus: A living thing that is somewhat like a plant, but doesn't make its own food, have leaves, or have a green color.

insect: A small, often winged, animal with six legs and three main body parts.

inspect: To look over something carefully.

larvae: Bugs in an early life stage that have a wormlike form.

mob: An informal term for a gathering of people or animals.

quadrillion: A number equal to 1 followed by 15 zeros.

vibration: A rapid movement back and forth.

FOR MORE INFORMATION

BOOKS

Amstutz, Lisa J. *Fast Facts About Ants*. Oxford, UK: Raintree, 2022.

Anthony, William. *Ants*. New York, NY: KidHaven Publishing, 2022.

WEBSITES

Ant Facts for Kids
www.pestworldforkids.org/pest-guide/ants/
Learn more about ants, and why they're more than just pests!

10 Cool Facts About Ants
www.natgeokids.com/uk/discover/animals/insects/ant-facts/
Discover more fun facts about ants from around the world!

Publisher's note to educators and parents: Our editors have carefully reviewed these websites to ensure that they are suitable for students. Many websites change frequently, however, and we cannot guarantee that a site's future contents will continue to meet our high standards of quality and educational value. Be advised that students should be closely supervised whenever they access the internet.

INDEX